which one doesn't belong?

A SHAPES BOOK / CHRISTOPHER DANIELSON

Also available: *Which One Doesn't Belong?* Teacher's Guide
For more information, visit: https://www.stenhouse.com/content/which-one-doesnt-belong

STENHOUSE PUBLISHERS
PORTLAND, MAINE

Stenhouse Publishers
www.stenhouse.com

Library of Congress Cataloging-in-Publication Data is available.

Book design by Tom Morgan (www.bluedes.com)

Manufactured in the United States of America

22 21 20 19 18   9 8 7 6 5 4

This book is different
from many other books
about shapes.

Every page asks the
same question, and every
answer can be correct.

Turn the page to see
for yourself.

Look at the shapes on
the next page. There
are many ways they are
alike and different.

Pick out a shape that seems
different from the others.

Which one doesn't
belong? Why?

Did you choose the shape
in the lower left?

If you did, maybe it's because
this shape isn't colored in.

Did you choose the shape
in the lower right?

If you did, maybe it's because
this is the only shape that
looks like it's leaning over.

Or maybe you said that
this shape doesn't belong
because it has three sides,
and the others have four.

Some people choose
this shape because it's
the only square.

Other people say that this
shape doesn't belong because
its angles are the wrong size.

On every page of this book,
you can choose any shape and
say that it doesn't belong.

The important thing is
to have a reason why.

How is your shape different
from the others?

What if you had picked
a different shape?

While the question is
the same on every page,
some pages are more
challenging than others.

You may need to put
the book down to think
and come back later.

So when you're ready, turn
the page and decide which
one doesn't belong.

which one doesn't belong?

which one doesn't belong?

which one doesn't belong?

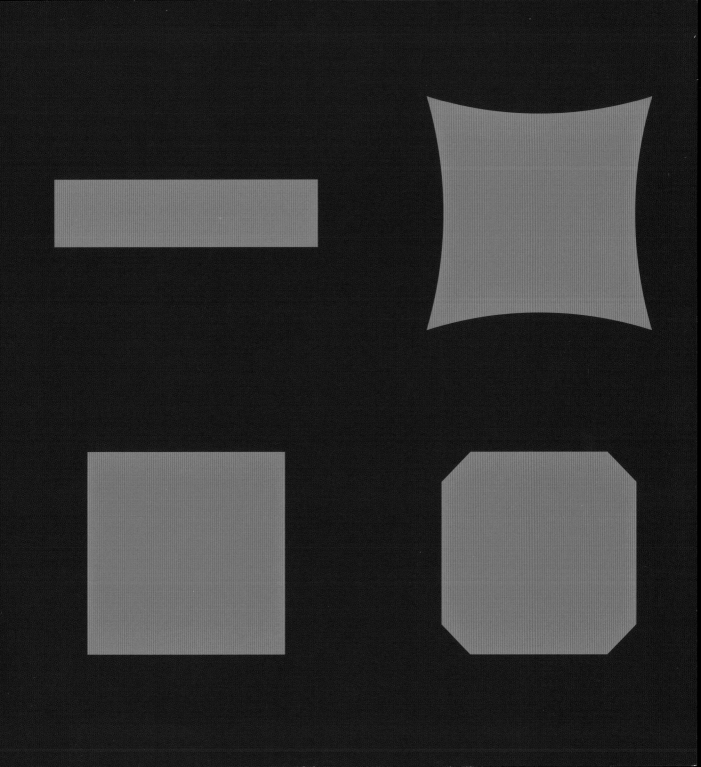

which one doesn't belong?

which one doesn't belong?

which one doesn't belong?

which one doesn't belong?

which one doesn't belong?

which one doesn't belong?

which one doesn't belong?

I made this book to spark conversations, thinking, and wonder.

I hope you will see similarities and differences in unexpected places.

I hope this is a book you will leave open, think about, and return to. I hope you will share it with others.

I hope you will send me your own sets of shapes to challenge me to say which one doesn't belong.

Find me at
talkingmathwithyourkids.com.